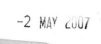

AS Geography
UNIT 2
2ND EDITION

AQA

Specification **A**

Module 2: Core Concepts in Human Geography

Philip Allan Updates
Market Place
Deddington
Oxfordshire
OX15 0SE

tel: 01869 338652
fax: 01869 337590
e-mail: sales@philipallan.co.uk
www.philipallan.co.uk

ISBN-13: 978-1-84489-028-6
ISBN-10: 1-84489-028-7

This Guide has been written specifically to support students preparing for the AQA Specification A AS Geography Unit 2 examination. The content has been neither approved nor endorsed by AQA and remains the sole responsibility of the author.

Printed by MPG Books, Bodmin

Environmental information
The paper on which this title is printed is sourced from managed, sustainable forests.

P00523

Contents

Introduction

■ ■ ■

Content Guidance

■ ■ ■

Questions and Answers

Introduction

About this guide

This guide is for students following the AQA Specification A AS Geography course. It aims to guide you through Unit 2, which examines the content of **Module 2: Core Concepts in Human Geography**.

This guide will clarify:
- the content of the module so that you know and understand what you have to learn
- the nature of the unit test
- the geographical skills and techniques that you will need to know for the assessment
- the standards you will need to reach to achieve a particular grade
- the examination techniques you will require to improve your performance and maximise your achievement

This introduction describes the structure of AS Geography and outlines the aims of Module 2. It then provides an explanation of some of the key command words used in examination papers. There is also advice concerning geographical skills, and learning and revision techniques.

The Content Guidance section summarises the essential information of Module 2. It is designed to make you aware of the material that has to be covered and learnt. In particular, the meaning of key terms is made clear.

The Question and Answer section provides sample questions and candidate responses at C-grade level and at A-grade level. Each answer is followed by a detailed examiner's response. It is suggested that you read through the relevant topic area in the Content Guidance section before attempting a question from the Question and Answer section, and only read the specimen answers after you have tackled the question yourself.

AS Geography

The AS is a new examination. It has been established as an intermediate standard between GCSE and A-level, the standard being defined as that which candidates have reached at the end of one year of study of an A-level course. Candidates may decide to continue to study for another year to take the A-level, which will be maintained at the traditional A-level standard. The AS is therefore a free–standing qualification but is also part of A-level.

All students taking the AS Geography examination for this specification (syllabus) must study Module 2 (which is one of three compulsory modules) and take the examination at the appropriate time. It is one of the six modules that contribute to the full A-level qualification. The structure of the qualification and the unit weightings for AQA Specification A are set out below.

AS

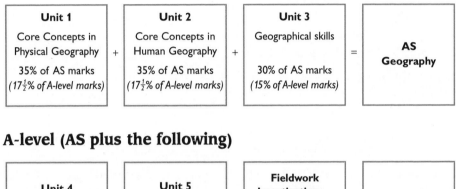

Unit 1		Unit 2		Unit 3		
Core Concepts in Physical Geography	+	Core Concepts in Human Geography	+	Geographical skills	=	AS Geography
35% of AS marks $(17\frac{1}{2}\%$ of A-level marks)		35% of AS marks $(17\frac{1}{2}\%$ of A-level marks)		30% of AS marks $(15\%$ of A-level marks)		

A-level (AS plus the following)

Unit 4		Unit 5		Fieldwork Investigation — Unit 6		
Challenge and Change in the Physical Environment	+	Challenge and Change in the Human Environment	+	Coursework or Unit 7 Written Alternative	=	A-level Geography
15% of marks		15% of marks		20% of marks		

The aims of Module 2

This second module of the AS course aims to help you:
- learn and apply knowledge and understanding of processes in human geography
- understand the influence people have on these processes and how the processes affect people
- apply this knowledge and understanding at at least two scales
- develop an understanding of the relationships between people and their environments
- learn and apply geographical skills
- understand that geography is dynamic
- reflect the importance of people's values and attitudes to issues and questions

Understanding these aims will be even more important to those of you who wish to continue on to A-level. A-level includes the requirement for candidates to be aware of 'synopticity'. This means that you must be able to apply your knowledge and understanding to demonstrate the links between the different parts of the geography specification. So the modules you study at AS are relevant to your A-level answers and should not be forgotten or discarded!

Examination skills

Before looking at typical examination questions and responses in the Question and Answer section, we will examine the broader skills that are essential for success in the examination. These fall into two areas: understanding the command words; and making the most effective use of the examination paper.

The importance of command words

Command words are used by the examiners to tell you what to do in order to answer examination questions effectively. The words are set out below, in an approximate order of difficulty and skill level, with an explanation of what they mean.

Command word(s)	Meaning
Describe...	State simply what is requested. Explanation or further comment is not required.
Name/State...	Identify briefly. One word may be adequate, but it may be better to use a sentence if in any doubt.
Distinguish between...	Define and state the differences between. Linking terms, such as 'whereas' or 'on the other hand', are essential.
Outline...	Describe, with a specific focus, the geographical element requested. For example, 'Outline the main features of ...' has more of a focus than 'Describe the main features of ...'
Outline the reasons for...	Give reasons for, with a specific focus, the geographical element required. The response will be briefer than a full explanation.
Account for/ Explain/Why...?	Give reasons for. The marks will be awarded for these reasons, rather than for description.
Give reason(s) for...	Some explanation must be offered.
Describe and explain...	Both elements, description and explanation, must be present for full marks. Ensure that examples of the mentioned theme are used in the response.
Compare...	What are the similarities between? Some element of contrast may be present.
Contrast...	What are the differences between? Two separate accounts will not meet the needs of this command; there must be a specific contrast or distinction between the elements.
Examine...	Give an overview of the elements which affect the theme, i.e. outline *and* explain.
Assess/To what extent...?	This requires an assessment of the importance of the factors involved in the response. This would be in an extended prose answer, rather than a short one.

The unit test

Module 2 is assessed by Unit 2. You have 1 hour to answer this paper, which accounts for 35% of the AS Geography qualification.

There are three stimulus–response questions, one from each of the three elements: (1) Population dynamics, (2) Settlement processes and patterns, and (3) Economic activity. You have to answer all the questions.

Each question is marked out of 20, so there are 60 marks for the whole paper. You are advised to spend 20 minutes on each question. The spacing on the questions allows two lines per mark for your response, which is written in the question book. It is therefore useful to allocate your time accordingly. The highest number of marks (10) is for the last part, so you should allow half the time on each question for this — about 10 minutes. This part requires a longer response in more continuous prose. The examiners are looking for good organisation in these answers, especially in the final part. Remember to check the command words so that you present the appropriate responses. Try to keep within the lines allocated, but if this proves to be impossible, use the lined pages at the end of the question booklet, making sure that you have clearly indicated what you have done and have identified each response.

Response levels

The first part of each stimulus–response question is given 1 or 2 marks for each point, while the final two parts are marked according to Levels. There are criteria for reaching these Levels depending on the quality of the geographical content and the use of geographical and English language. These parts require a fuller development of geographical understanding and language skills.

Part (b) of each question is marked in two Levels; part (c) in three Levels. These are as follows:
- Level 1 (basic) responses offer one or two points without examples.
- Level 2 (clear) responses show some understanding, better use of language and make points with examples.
- Level 3 (detailed) responses show detailed understanding, make several points with examples, and use language effectively.

The command words are important, so make sure you remember to follow the instructions.

Geographical skills

As an integral part of your studies for this unit, you are required to develop a variety of skills. There are six types of geographical skill specified at AS:
- basic skills
- graphical skills
- cartographic skills

- ICT (information and communication technology) skills
- statistical skills
- investigative skills

Basic skills

Base maps

The drawing or use of photocopied base maps is essential for settlement (land use) studies, and the use of overlays for delimitation of the central business district (CBD). Annotation is an effective addition to such maps.

Sketch maps

These require you to select only relevant information to produce, for example, sketch maps of the location of industry, population distribution or migration patterns.

Atlas

The use of an atlas is essential so that global distribution of population, megacities and industrial locations can be identified. The atlas can also provide the basis for sketch maps.

Photographs

These are effective as an aid to interpretation, for example photographs taken of CBDs in the field. Other uses include the impact of de-industrialisation and the characteristics of different land-use in cities.

Literacy

You need to develop literacy skills during the AS course. The assessment units require the ability to respond to both short-answer and extended prose questions.

Graphical skills

You should be familiar with the majority of these skills, but some will be new to you. You are expected to be able to interpret and construct the following.

Line graphs

Four types are specified:
- **simple line graphs**, for example world population growth
- **comparative line graphs** — two line graphs on the same axes for comparison, for example the demographic transition model
- **compound line graphs** — line graphs with amounts totalled, for example world population growth, with each continent shown
- **divergent line graphs** — line graphs showing variation from a mean or other fixed point, which is the central horizontal axis, for example variations from the mean in rural and urban populations between countries

Bar graphs

Four types are specified:
- **simple bar graphs**, for example industrial output

- **comparative bar graphs** — bars side by side for comparison, for example urban population growth in MEDCs and LEDCs
- **compound (or divided) bar graphs** — bars divided into their components, for example the proportions of retail warehouses, town centres and out-of-town centres
- **divergent bar graphs** — bar graphs showing variation from a mean or other fixed point which is the central horizontal axis, for example variations from the mean population by type of area

Scattergraphs — and the use of the best fit line

These plot the relationship between the point data for two variables, for example the number of functions against city size.

The best fit line is the average straight line passing between the points — the closer the points to the line, the closer the relationship between the variables. The steeper the gradient, the more rapid the change in the relationship.

There are three relationships of which you should be aware:
- **positive** — the two variables increase or decrease together and the trend of the line rises to the right, for example number of functions and city size
- **negative** — one variable increases while the other decreases and the trend of the line falls to the right, for example city size and rank
- **no relationship** — there is no discernible trend to the pattern (this can indicate clustered data instead)

You need to be aware of the use of log–log and log–normal graph paper for scatter-graphs.

Pie charts

Pie charts use sectors of a circle to show proportions of a whole, for example the proportions of contribution to a country's economy by various sectors of economic activity.

Triangular graphs

These are used to show the proportions of three contributing elements, such as primary, secondary and tertiary sectors' contributions to a country's economy.

Lorenz curves

These show inequalities in distributions. An even distribution is shown by a straight line at 45°; the further the actual distribution is from this line, the more varied it is. These are most frequently drawn to show variation in population distribution by continent.

Pyramids

These are used to show amounts by groups. Population pyramids are the most common example, showing age/sex of groups.

Logarithmic scales

You need to have an understanding of logarithmic scales. These are used to plot rates of change.

Cartographic skills

Ordnance Survey maps

You need to be familiar with these at a number of scales: 1:50 000, 1:25 000, 1:10 000 and 1:1250. You will be familiar with the smaller-scale maps from your GCSE studies, but the larger-scale maps/plans are of great value for work involving settlement studies. If these maps are not available in your centre, they are held in the reference sections of larger public libraries and in the planning departments of local authorities.

Choropleth (shading) and isoline maps

These two types of map are available for interpretation and also can be constructed from your own data.

Choropleth maps use shading to show the spatial distribution of a factor, dividing data into ranges of values represented by a different colour or texture of shading, for example population density in a city by ward.

Isoline maps join points of equal value, for example journey times to regional shopping centres. Isoline maps show spatial changes in patterns of data.

Remember that choropleth maps should only be produced for data in the form of ratios or density and not simple numbers, whereas isolines can be used for both types of data.

Goad maps

These are specialist maps of land-use functions for most urban centres in the UK. Many are updated annually and therefore provide a snapshot of land-use in central business districts. Comparisons with previous editions can provide valuable information on changes to central areas.

Many of the graphical skills covered in the previous section, such as bar graphs and pie charts, can be demonstrated on a base map to add greater variety to both types of skill and show ability at higher level. For example, pie charts could be used on a base map to show the varying proportions of the three main sectors of the economy (primary, secondary and tertiary) for a number of countries.

ICT (information and communication technology) skills

Photographs

You need to be familiar with ground and oblique aerial photographs and be able to interpret the geographical features they depict through written answers and sketching. Housing characteristics and industrial landscapes can be identified using photographs.

Satellite images

The spread of urban and industrial areas can be identified from satellite images.

Databases

Population/census data and industrial statistics are accessible.

Internet

This is an expanding source of information for geographers, offering an array of databases, case studies and news sources. You will not be specifically examined on this skill, though reference could be made to it. Information is available, for example, on multinationals, population changes and urban development.

Video and television programmes

You will not be specifically examined on this skill, though reference could be made to these media. Issues of urban expansion and industrial change are covered.

Statistical skills

Measures of central tendency

The **mean** is the *average* value of a set of data (arithmetic mean).

The **mode** is the *most frequently occurring* value in the data set.

The **median** is the *middle value* of the data set.

Means of dispersion

You will need to know how to plot and interpret a **dispersion diagram**. This shows the range of data covered and permits a fuller understanding of the interquartile range and standard deviation.

The **interquartile range** is the range between the 25th and 75th percentage points in the data set. For example, if the data listed population growth rates for 100 countries, then the interquartile range would encompass the 25th highest to the 75th highest level, i.e. 25 points either side of the median. It shows the dispersion of the data from the median and thus shows the variability of population growth rates.

The **standard deviation (SD)** is a statistical means of assessing the average amount by which the distribution of data varies from the mean. For example, taking the situation mentioned above, the population growth rate might be 2.0%. An SD of 0.2% would indicate that population growth rates tend to have little variation from country to country, whereas an SD of 0.5% would indicate a wider variation, as illustrated by the following table:

Mean = 2.0%	SD = 0.2%	SD = 0.5%
Within plus or minus 1 SD (68% of values)	Between 2.2% and 1.8%	Between 2.5% and 1.5%
Within plus or minus 2 SD (95% of values)	Between 2.4% and 1.6%	Between 3.0% and 1.0%
Within plus or minus 3 SD (99% of values)	Between 2.6% and 1.4%	Between 3.5% and 0.5%

In terms of the examination, you might be expected to be aware of the skill, what the results mean and to be able to carry out some simple stages of the calculations.

Correlation tests

Spearman's rank correlation coefficient is the only correlation test you need to know. This is a relatively straightforward test to establish whether a relationship exists between two sets of ranked data and what type of relationship that might be. It is used for data shown on a scattergraph (see pp. 9–10), i.e. with positive or negative relationships. The range of results is always within the range of +1.0 (perfect positive correlation) to –1.0 (perfect negative correlation). It follows that a result closer to 0 indicates a very weak relationship (more like a clustered distribution). The minimum number of pairs of data is 6 (according to statisticians), though most geographers would expect a minimum of 10 pairs to ensure a greater reliability of the result.

This test is very frequently used for population work, including population growth rates correlated with economic growth rates and, for retail studies, distance travelled correlated with frequency of journey.

In terms of the examination, you might be expected to be aware of the skill, what the results mean and to be able to carry out some simple stages of the calculations. You will not be expected to use significance at AS.

Investigative skills

The identification of geographical questions and issues

Unit 2 provides a wide range of opportunities for investigative study. Settlement studies are the main focus of much fieldwork, though population and industry also provide opportunities. Examples of some simple hypotheses include 'Migration is age-related', 'Land-use conforms to Robson's model' and 'The number of retail functions is related to the size of the centre'.

The selection of relevant primary and secondary data and an assessment of their validity

The popularity of settlement-related fieldwork means that appropriate data can be collected. The nature of the hypotheses to be tested will define the actual data collected, e.g. land use, land values, questionnaires and so on. Secondary data on settlements, for example, are available from a number of sources, including the Office of Population Censuses and Surveys, and local authorities. The Internet is a valuable means of accessing such data.

You also need to understand the reliability of the methods of data collection employed and of the data collected.

The processing, presentation, analysis and interpretation of the evidence collected

These skills are developed as the course progresses and are not confined to data collected in the field. They can be assessed in a general sense in questions in Units 1 and 2 and will not be formally assessed in Unit 3. Writing up fieldwork and analysis of data downloaded from the Internet will develop such skills. In terms of population settlement, retailing and economic activity, a comparison of the results, interpreted in relation to theory, will fulfil the requirements of this section.

The ability to draw conclusions and show an awareness of their validity

This skill is again not confined to data collected in the field. It could be assessed in the general sense when responding to questions in Units 1 and 2, and will not be formally assessed in Unit 3. A conclusion to and an evaluation of the success of any geographical fieldwork exercise would be appropriate here.

The awareness of risks when undertaking fieldwork

This is essential. It is not advisable to undertake fieldwork on your own, for example. Assistance from others is needed to collect the data and to ensure your safety in any geographical fieldwork situation.

Investigative skills are best developed by a programme of fieldwork undertaken in the AS year. Preparation for work in the field, the collection of data and their interpretation and evaluation are demonstrated clearly by writing up the fieldwork in the format suggested by the board. Your teachers will be able to advise you about this.

Techniques for learning and revision

Most of you will be taking the AS examination at the end of, or during, a one-year course, with the examination taken in January or June. This means that there is no surplus time available for teaching the subject content. You must ensure that, from the start of the course, you establish good working practices to make the most of the time available.

- It is important that you do not fall behind with work during the year. New material will be taught each week so, if you are unavoidably absent (because of illness, for example), do make sure you are able to make up the missed work as quickly as possible.
- You will probably have a steady stream of homework during the course. This is likely to take a variety of forms, ranging from working from your textbook or other sources, to practising examination questions.
- Read widely from a variety of sources. Television programmes are also relevant. The information you gather will enable you to develop a number of case studies for use in your examination answers.

If you keep on top of the work, your revision programme will be more relevant and straightforward in the lead-up to exams.

The specification is divided up into modules, as we have already seen. Each module is divided into three elements and each of these into three or four sub-elements. **Module 2: Core Concepts in Human Geography** is divided as follows:

Elements	Sub-elements
Population dynamics	Population change Migration Population structure
Settlement processes and patterns	Urbanisation and suburbanisation Counter- and re-urbanisation Size and spacing of settlements
Economic activity	Secondary activities Tertiary and other activities

Revision can be more easily structured by taking the sub-elements and focusing on them. Note that it is better to revise the sub-elements in the order in which they appear, or there might be the risk that points will not make sense!

Some tips on revision

- Having selected a topic for revision, read and learn the material you have for this topic, e.g. notes, handouts, worksheets etc.
- Refer to your textbooks and to this publication. You might also find Raw, M. (2000) *AS/A-level Geography Exam Revision Notes* (Philip Allan Updates) a useful guide.
- Learn the relevant case studies. For AS you probably need no more than two for each element/sub-element and these should be at different scales to meet the examination requirements.
- Practise sample questions, keeping to the appropriate timings. Use the questions in the last section of this guide for this purpose, taking care not to look at the sample answers and examiner's comments until you have attempted the questions. There are other specimen questions available, so consult your teacher/lecturer for advice.
- Apply your knowledge and understanding when practising so that your answers reflect the demands of the question.
- Allow yourself adequate time for revision. Little and often is usually better than concentrated pressure at the last minute.

Content Guidance

There are three elements in the specification content for Module 2:

(1) Population dynamics

(2) Settlement processes and patterns

(3) Economic activity

In this section, the key concepts of each of these topics are explained, together with a breakdown of what you need to know and learn.

Details of case studies are indicated under the sub-elements. One case study per sub-element is required and the scales are indicated by the following letters:

(S) Small scale (could be fieldwork)
(R) Regional scale (smaller scale than national)
(N) National scale
(G) Global scale

Points to note

- The same cases can be used for more than one sub-element, as appropriate.
- In a small number of cases, more than one case study is specified. This is indicated at the appropriate point using the code letters above. The most frequent will be comparisons of LEDW and MEDW examples.

Population dynamics

Population change (N)

Components

There are four components of population change:
- births
- deaths
- immigration
- emigration

Natural population change for an area is the birth rate minus the death rate.
- The **crude birth rate** is the number of live births per 1000 population per annum.
- The **crude death rate** is the number of deaths per 1000 population per annum.

The crude birth rate minus the crude death rate gives the natural increase (or decrease) in population, and is expressed as a percentage. For example:

Crude birth rate (per 1000, per annum)	Crude death rate (per 1000, per annum)	Natural increase (decrease) in population (%)
15.0	12.0	0.3
11.0	13.0	(0.2)

Population change can also be measured by fertility rates and infant mortality rates.
- The **general fertility rate** is defined as the number of live births divided by the number of women of childbearing age (15–49). It can also be expressed as the number of children born per woman.
- The **infant mortality rate** is the number deaths of infants under 1 year old per 1000 live births. This is used as an indicator of the level of economic development of a country, because the infant mortality rate falls with greater levels of economic development.

You need to:
- be aware of the components of population change
- be able to define crude birth and death rates
- be able to calculate and understand population growth (and decrease) rates
- be able to define fertility and infant mortality rates and to understand their significance

The demographic transition model

The interaction of crude birth and death rates causes changes in a country's population over time. These rates change for a number of reasons — economic, social, environmental and political. The demographic transition model (DTM), shown below, provides a vehicle for the study of such changes in both LEDCs and MEDCs.

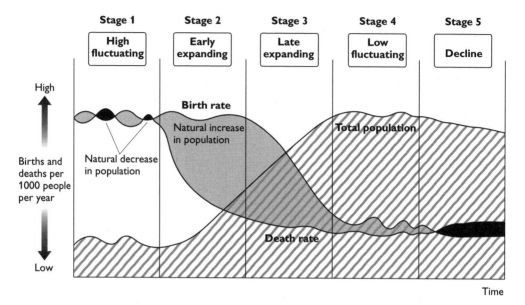

Stage 1

High fluctuating stage. Birth and death rates are over 35/1000 and fluctuate. Population growth is small. Reasons for high birth rates include:
- no birth control/family planning
- high infant mortality rates, encouraging more children to be born
- children are a source of income at an early age
- children are a sign of virility and fertility
- some religions/cultures encourage large families

Reasons for high death rates include:
- disease and plague
- famine
- poor nutrition
- poor hygiene
- undeveloped medical science

Stage 2

Early expanding stage. Birth rates high, death rates falling to 20/1000. Population increases rapidly. Reasons for falling death rates include:
- improved public health
- better nutrition
- lower child mortality
- improved medical facilities

Stage 3

Late expanding stage. Birth rates fall rapidly to 20/1000, death rates to 15/1000. Population increases slowly. Reasons for falling birth rates include:

- changing socio-economic roles of women
- smaller families preferred
- fashion
- increased wealth
- compulsory schooling makes children expensive
- greater access to education for women
- lower infant mortality rate
- availability of family planning and the role of government

Stage 4

Low fluctuating stage. Birth and death rates are low (16/1000 and 12/1000 respectively). There are increasingly low fertility rates.

Stage 5

The birth rate falls below the death rate, causing a population decline. This stage is not in the original model, but has been observed in some west European countries. It could in the long term be part of the normal fluctuation within stage 4.

You need to:

- be able to name the stages of the DTM
- be able to explain the reasons for the changes to birth and death rates
- be able to construct or complete an annotated diagram of the model

Uses of the DTM:

- Universality — all countries can be placed on the model.
- It provides a starting point for the study of demographic change over time.
- The timescales are flexible.
- It is easy to understand.
- It enables demographic comparisons to be made between countries.

Limitations of the DTM:

- The lack of the fifth stage in the original model.
- It is Eurocentric, assuming that all countries will follow the European stages in the same timescale. This is not the case for NICs (newly industrialising countries), which are changing more rapidly.
- It assumes that the same socio-economic changes that took place in Europe will occur in other countries for the same reasons.
- The role of governments is not covered.
- The contribution of migration to population change is not covered.

You need to:

- be aware of the usefulness and the limitations of the DTM
- be able to apply the DTM to one case from both the LEDW and the MEDW (N)

Migration (R and/or N)

Definition and types

Migration is a permanent or semi-permanent change of residence. (Commuting on a daily basis is called circulation and is not generally included as migration.) The types of migration can be classified in a number of ways.

Factor	Types
Timescale	Permanent, semi-permanent, seasonal
Motivation	Forced, voluntary, economic, political, social, retirement
Distance	Internal, external (international)
Source/destination	Rural–urban, urban–rural, urban–urban, rural–rural
Settlement processes	Urbanisation, suburbanisation, counter-urbanisation, re-urbanisation

You need to:
- be able to define migration
- be aware of different types of migration
- be able to apply these types to one case from the MEDW and one from the LEDW

Models of migration

A number of models and concepts are specified:
- **Push–pull (Carr)** — people migrate as a result of negative factors which push them away from the area in which they live and positive factors which pull them to the new area.
- **Intervening obstacles (Lee)** — the decision to migrate depends on the number of obstacles (negative factors) between the origin and the destination. The greater the number of obstacles, such as mountain barriers or international boundaries, the less likely it is that migration will occur.

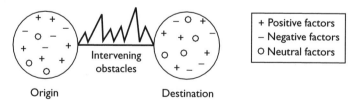

Origin   Intervening obstacles   Destination

+ Positive factors
− Negative factors
O Neutral factors

- **Intervening opportunities (Stouffer)** — the decision to migrate is directly related to the number of intervening opportunities (positive factors), and inversely related to the number of intervening obstacles.
- **Distance decay** — the number of migrants declines inversely with distance.

You need to:
- know and understand the push–pull and Lee's models of migration
- be able to apply these models to the case studies specified

Migration consequences

You need to be aware of the consequences of migration, including the economic, environmental, social and cultural impacts, as well as the attitudes and values relating to migration. These generalised consequences can be applied to the areas of both origin and destination.

Impact	Origin	Destination
Social	Younger age groups migrate (20–34), leaving an older population	Younger age groups increase proportion of the population
	Males more likely to migrate, decrease in proportion	Males increase in proportion
	Birth rate might fall; death rate will rise	Birth rate might rise; death rate will fall
	Marriage rates will fall; family structures break down	Marriage rates will rise
Economic	Those with skills and education leave, causing labour shortage or the advantage of reduced pressure on resources	Those with skills and education arrive, causing labour surplus and new drive to the economy; take-up of menial jobs
	Dependence on remittances — gain to economy	Loss of remittances to the economy
	Migrants bring new skills back on return	Skills exported on leaving
Environmental	Decline of farming; land abandoned	Pressure on resources; temporary housing and shanty towns; pollution; poor public health etc.
Cultural	Loss of males and young families later causes a loss of cultural leadership and tradition	Arrival of a new group of people can cause friction, especially if cultural identity is retained; the attitudes of local people to new migrants, and vice versa, should be studied in the light of your own attitudes and values
		New foods, music, etc. introduced

You need to:

- be aware of and understand the consequences of migration on areas of both origin and destination
- identify and understand the attitudes and values of people in areas of origin and destination

Case studies at various scales

Migration can be studied at a number of scales. You are required to study one at a national scale and one at an international scale. A number of influences can be identified in each case, including physical, economic, social, cultural and political factors.

If the UK is selected as the case study, you will be able to look at changes over time, both internally and externally, although other cases will involve similar changes. One case must involve international migration.

Scale	Possible case study
Small scale	Migration patterns within a settlement local to you. This could be a village, town or other urban area, and data could be collected by fieldwork
UK	International migration (both emigration and immigration over the last few decades) and regional movement since 1945
EU	Although there are other examples (West Indies to the UK, North Africa to France), the case of the migration of Turks into Germany is well documented and also covers two scales
LEDCs	See EU above

You need to:

- apply and understand two case studies of economic and refugee migration — one at a national scale and one at an international scale.

Population structure (N)

Population structure is the division of a population by age and sex. This can be shown by means of a population (or age–sex) pyramid which has the population in five-year age bands on the vertical scale and the number or percentage of males and females on the horizontal scale.

The pyramid also shows the **life expectancy** of the population by its height. Life expectancy is the average lifespan of a person born in a specific year. It is greater for females. For example, in the UK the life expectancy in 1999 was 74 for men and 80 for women. In most countries of the world, life expectancy is rising as standards of living rise, but it has fallen in certain countries, such as Russia, because of falling living standards.

You need to:
- be able to define population structure
- be aware of the make-up and construction of population pyramids
- be able to define life expectancy and explain the reasons for change

Population pyramids

Population pyramids can show the following:
- the result of births minus deaths in specific age groups
- the effects of migration
- the effects of events such as war, famine and disease
- the overall life expectancy and the dependency ratio

The **dependency ratio** is the proportion of the population that is economically non-productive compared with the proportion that is economically productive — that is, the proportion aged 0–14 (non-working) and 65 and over (retired) compared with the proportion aged 15–64 (working). The higher the dependency ratio, the more the non-productive proportion is dependent on the productive.

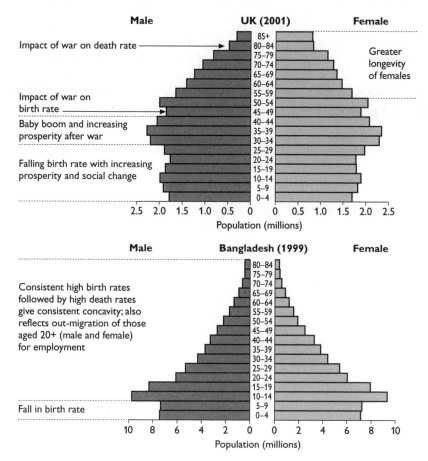

Pyramids and the DTM

The DTM can be used as a vehicle to demonstrate changes in population structure spatially and over time. This is clearly seen by the characteristic shapes and names of the pyramids at each stage of the DTM.

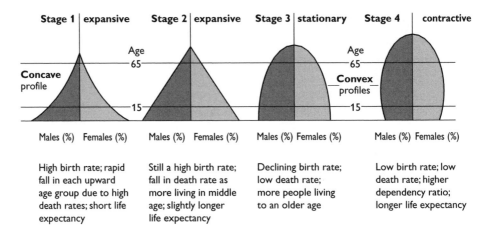

High birth rate; rapid fall in each upward age group due to high death rates; short life expectancy	Still a high birth rate; fall in death rate as more living in middle age; slightly longer life expectancy	Declining birth rate; low death rate; more people living to an older age	Low birth rate; low death rate; higher dependency ratio; longer life expectancy

These changes can be applied over time. More developed countries, such as the UK, would have passed through each of these stages to reach the current one.

Consequences of change in population structure: ageing and youthful populations

Changes in population structure have economic, demographic, social and political implications. Ageing populations are a consequence of the decline in birth rates that most countries have experienced or are experiencing, and tend to be important in MEDCs such as the UK. The structural type is known as contractive/regressive.

The contractive/regressive population structure

The characteristics of this structural type are:

- an ageing population
- low birth and death rates
- a narrowing base to the pyramid
- a smaller proportion of younger age groups
- a larger proportion of older age groups
- longer life expectancy
- more females in older age groups
- more males in younger age groups

This structure has many implications for the population.

Economic implications:

- increasing dependency ratio, despite the lower birth rate, as life expectancy rises and young people enter the labour market later in life

- costs of care for the elderly rise (medical care, hospitals) as more elderly survive and medical inflation is greater — pensions take a higher proportion of government expenditure as life expectancy rises
- costs for education (especially higher education) rise as school leaving age effectively rises
- a decline in services for the young, but an increase in economic opportunity for services for the elderly, e.g. sheltered housing

Demographic implications:
- a smaller proportion of younger age groups
- a larger proportion of older age groups
- longer life expectancy
- more females in older age groups
- more males in younger age groups

Social implications:
- more elderly people living together in sheltered housing or coastal towns
- social interaction might be restricted and the social mix unbalanced

Political implications:
- greater political power for older voters as their numbers increase
- higher expectations of the older age group can only be met if the working population is prepared to meet the costs of supporting them

In the UK there has been great controversy in recent years as the political parties and the electorate fail to agree on the levels of taxation required to provide the services demanded.

Youthful population structures occur in LEDCs, such as Bangladesh. This structural type is known as expanding/progressive.

The expanding/progressive population structure
The characteristics of this structural type are:
- a youthful population
- high birth rates and falling death rates
- a broadening base to the pyramid
- a large proportion of younger age groups
- a smaller but increasing proportion of older age groups
- low but rising life expectancy
- more males in the younger age groups

This structure has many implications for the population.

Economic implications
- the large proportion of young people demanding provision of education and health services
- lack of time to adapt to the rapid population increase
- large supply of young, active labour into the economy
- dynamic, innovative labour supply

- increasing demand for services, expanding economy
- low dependency ratio on reaching the age of economic activity
- higher rates of unemployment and underemployment

Demographic implications
- larger proportion of young population
- smaller proportion of older age groups
- more males in younger age groups

Social implications
- inadequate provision of schools and health services
- services available mainly in urban areas

Political implications
- young, dynamic political class created
- increasing numbers of voters: a force for change

You need to:
- be aware of the economic, demographic, social and political implications of an ageing population in the UK and a youthful population in an LEDC example
- explore your own values in this context

Settlement processes and patterns

Urbanisation and suburbanisation (S or R/N)

Urbanisation is defined as an increasing proportion of a country's population living in urban areas. There have been two main waves of urbanisation:
- in MEDCs during and following the industrial revolution — as a result of economic development
- in LEDCs since 1950 — as a result of migration from rural areas and high rates of population growth in urban areas

Suburbanisation is defined as the movement of people from the central and inner areas of a city to the surrounding residential areas. It has taken place in MEDCs as transport has developed, enabling people to commute to their place of work, and is characterised by lower building densities.

You need to:
- be able to define the terms 'urbanisation' and 'suburbanisation'
- know where and when the two processes have occurred

The growth of settlement: MEDCs

Accelerated urban growth took place in MEDCs as a result of industrialisation, which encouraged the migration of potential workers from rural areas to meet the demand for labour in factories and mines. Housing was built near the factories and mines, as transport was poorly developed and wages low. Building was at a high density and conditions were poor. Over time, because of legislation and the increasing wealth of workers, housing conditions improved. In the UK, the typical late nineteenth-century terraced housing was the result of these developments. This housing spread away from the centre of the urban areas to form what is now recognised as the inner city.

You need to:
- know and understand the reasons for accelerated urban growth in MEDCs

The growth of settlement: LEDCs

Accelerated urban growth in LEDCs has been spectacular in recent years, resulting in megacities (very large cities). The fastest-growing are in Latin America, southeast Asia, Japan and Korea, with Mexico City, São Paulo, Tokyo, Calcutta and Bombay all with over 16 million inhabitants in 2000.

The reasons for this growth include rural–urban migration (push–pull) and natural increase, without the same degree of industrialisation that took place in MEDCs. The wealthy tend to live close to the city centre and the poor tend to live further away. Consequently, the quality of housing declines away from the centre, the reverse of the situation in MEDCs.

Million cities (cities with populations of 1 million or above) first appeared in MEDCs, but the majority of recent ones are in LEDCs.

You need to:
- know the areas of the world experiencing accelerated urban growth
- know and understand the reasons for accelerated urban growth in LEDCs
- be able to account for the changing distribution of million cities over time

Consequences of accelerated urban growth

The building of shanty towns (*favelas* in Brazil; *bustees* in Calcutta) is a direct conse-quence of rapid urban growth, as there is not enough housing to accommodate all of the population. On average, 30% of an LEDC city's population live in such settlements. The houses are constructed from any available material.

The following factors are associated with shanties:
- services are poor, with little running water, mains drainage or rubbish collection
- the streets are frequently open sewers and carry flood water when rains occur
- they have limited electrical power and a lack of schools, teachers, hospitals, doctors and nurses

- there is frequent pollution of drinking water, leading to the spread of infection and disease (typhoid, cholera and dysentery)
- the air, land and water are often polluted by industry, resulting in high infant mortality rates and low life expectancy
- most people are unemployed or underemployed and find work in the informal sector of the economy
- transport is poorly developed and roads are unable to cope with the volume of traffic

However, in time many shanties develop to an adequate standard. Authorities provide water and power supplies and allow the inhabitants to gain title to the land. In the *periferia* (periphery) of São Paulo in Brazil, the 'site and services' policy (funded by the World Bank) has been successful and involves the local people helping to improve their own environment.

You need to:
- be able to describe and account for the consequences of urban growth
- be able to show how responses to urban growth have changed over time
- be aware of the disadvantages found in shanty towns, and the signs of improvement

Suburbanisation in MEDCs

Suburbanisation in MEDCs occurred for the most part in the mid to late twentieth century. This was a consequence of a number of factors including improvements in transport (the bus, electric train and motor car), increasing wealth, and availability of housing, employment and other facilities in the suburbs. The availability of lower-cost land assisted the construction of cheaper, yet better quality, housing, at a lower building density. The wealthier, middle-class inhabitants were able to travel to their place of work by car, train or bus, allowing the segregation of residential land and land used for other functions. For example, the population of Greater London doubled between 1921 and 1951 as a result of this process. In addition, there was segregation by social class as suburbanisation took place.

The construction of very large housing estates in the suburbs of major cities in the UK showed the decentralisation forces at work. Many of these housing estates were privately owned, but a significant number were built by local authorities to rehouse people from the inner cities. This caused greater social segregation. Examples of suburbanisation, for further study, can be found in any town or city.

Suburbanisation has been limited in the UK by the establishment of 'green belts' around urban areas. These have prevented further spread of residential areas and other developments. The suburbanisation process has continued in other parts of the MEDW where such planning restrictions are not found; the USA is the classic case.

You need to:
- understand the reasons for suburbanisation
- be aware of the consequences of the process in terms of functional and social segregation
- be aware of the reasons for the restrictions on suburbanisation imposed by planning controls in the UK

Counter-urbanisation and re-urbanisation (S)

Counter-urbanisation

Counter-urbanisation is the movement of people from an urban area into the surrounding rural area. This is frequently characterised by the development of commuter villages. It is distinguished from suburbanisation by the final location being outside the existing urban area.

Causes	Consequences
Accessibility — railways and motorways allow access to places of employment	Middle-class immigrants; social structure changes
Mobility — increasing levels of car ownership	Increased traffic/congestion; dependence on car
Increasing wealth — house prices and costs of travel become affordable	House prices rise beyond the reach of locals
Decline of farming — less labour needed and land might be run down	New estates of detached and semi-detached houses built; renovation of farm buildings
Out-migration of agricultural workforce — allows new migrants to move in	Young to middle-aged professional people move in, plus some of retirement age
Planning — existing urban margins are clear, causing green belt areas in which the process can occur	Light industry might develop in the area
Fashion — people wish to live in such areas	Shops might decline as a result of competition from supermarkets in sub-urban areas (other areas might have more retail services); more restaurants; primary schools might flourish or in some cases close; village society/organisations taken over by middle classes

Changes to rural settlements in the UK as a result of counter-urbanisation

Changes to rural settlements have occurred most strongly as a result of counter-urbanisation. (Rural settlements remote from urban areas show rural–urban migration, with urbanisation as the main process.) These changes have accelerated during the second half of the twentieth century.

Land-use patterns in suburbanised villages

Suburbanised villages have experienced much change in recent years. The influx of new population has been reflected in the changes to the land-use structure of the village. These changes have taken a number of forms, including new detached and semi-detached houses and bungalows, both on individual plots and on estates of varying sizes, and the conversion of old farm buildings. Compare your case study of a suburbanised village with the model below. (The processes that underlie the changes to the structure have been covered above.) The model summarises the types of change that have occurred in suburbanised villages.

1 Original village core
2 Infills, modifications and accretions
3 Ribbon development
4 Adjuncts
5 Isolates

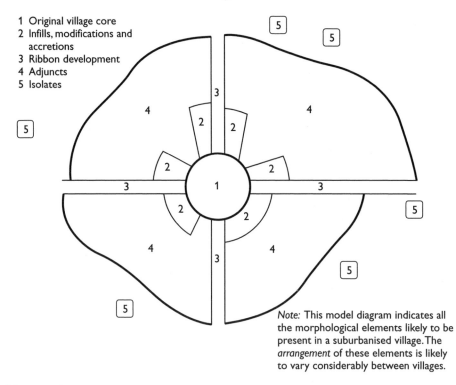

Note: This model diagram indicates all the morphological elements likely to be present in a suburbanised village. The *arrangement* of these elements is likely to vary considerably between villages.

You need to:
- know and understand the model of a suburbanised village
- consider the processes that underlie the changes in land-use patterns in suburbanised villages
- undertake a small-scale study of a suburbanised village

Re-urbanisation

Re-urbanisation is the movement of people into the centre or inner city as part of a process of urban regeneration. The migrants are frequently middle class and the process is also known as gentrification. In some cases the investment by central and local governments as part of a strategy for urban regeneration might result in a more varied social structure.

You need to:
- know and understand the reasons for counter-urbanisation
- know and understand the consequences of counter-urbanisation

- be able to define counter-urbanisation and re-urbanisation
- be able to distinguish between suburbanisation and counter-urbanisation
- apply this understanding to a case study of both processes in the UK

Factors favouring re-urbanisation: a case study

Re-urbanisation has been a feature of many inner and central areas of cities in recent decades. The regeneration of London's Docklands has taken place since 1980. The establishment of the London Docklands Development Corporation (LDDC) was the catalyst for major change. The Docklands area had been abandoned as shipping tended to use facilities down-river at Tilbury. The oldest and smallest docks were by Tower Bridge and the largest on the Isle of Dogs (the Royal group of docks).

The LDDC reclaimed over 600 hectares of derelict land and helped to improve the transport infrastructure. The Docklands Light Railway, extensions to the Jubilee Line, London City Airport and the Limehouse Link all improved access to the area. The public investment of over £1 billion was more than matched by over £8 billion of private investment. The flagship development at Canary Wharf helped to establish more than 40 000 jobs, though many local people were unable to gain employment.

The social regeneration of the area involved mostly private ownership, though some 'affordable' housing was built further away from the centre. The old warehouses alongside the river near the City were converted into expensive apartments and this encouraged restaurants, businesses and retailers into the area. In particular, the accessibility of the location for the City and Canary Wharf caused the area to become very fashionable. The apartments are frequently used as mid-week residences, with the owners returning to their homes in suburbanised villages at the weekend. The proportion of private housing increased from 5% to 44% and the resident population from 39 000 to 61 000.

Similar social change (though without such major regeneration) has taken place in inner areas such as Islington and Fulham, which became fashionable for the wealthy because of their access to the centre of London and the good quality but (at the time) relatively cheap housing that had been constructed in the nineteenth century. The poorer residents were pushed out of the area by the rise in house prices. Similar examples of re-urbanisation can be found in most urban areas around the UK.

You need to:
- be aware of the factors encouraging re-urbanisation
- be aware of the consequences of re-urbanisation
- apply these factors and consequences to a UK case study

Greenfield vs brownfield residential developments

From studying suburbanisation, counter-urbanisation and re-urbanisation it is clear that issues arise regarding the availability of sites for new residential development. The government in the UK has been advised to consider plans for new homes, particularly in the south and east of the country, as patterns of family structure change. To meet this projected demand, both brownfield and greenfield sites (defined below) will need to be used, but the values and attitudes of decision-makers are not necessarily those of the people who already live in these areas.

Brownfield sites	Greenfield sites
Derelict sites in urban areas	New sites, usually on agricultural land in green belts around urban areas
The land is available, but can be costly to reclaim if it has been polluted by industrial use; this information may not be readily available	Land is not available unless planning permission has been obtained; there will usually be a public enquiry and a delay of several years, adding to costs
Housing is likely to be built at a high density to reflect the cost of the land; there is less demand for such housing as it is in less fashionable areas	Housing will be at a relatively low density compared with the brownfield sites; there is great demand for such housing as it is in fashionable areas
Infrastructure is normally present, though existing facilities can become overloaded	Infrastructure costs are high as new sewerage, water, gas and electricity supplies have to be considered; similarly, new schools and medical facilities might be needed
Sites tend to be small patches of land	Sites tend to be large
The environment is improved	The environment is harmed
Existing services are close by	New service provision is required

You need to:
- be aware of the advantages and disadvantages of brownfield and greenfield sites for residential development
- be able to evaluate the issues surrounding such developments

Values and attitudes

The UK government's preferred option is to develop brownfield sites, and only to use greenfield where necessary. However, you need to consider whether people should be forced to live in brownfield sites if they do not wish to. In some cities, recently built local authority housing is being demolished as there is no demand for it.

On the other hand, should green belt land be used so that people can live where they wish? Is the damage to the environment necessary? Remember, not all housing can be built on brownfield sites if demand is to be met. You need to be able to appreciate the values and attitudes involved in this debate.

You need to:
- explore the values and attitudes of decision-makers as well as your own

Size and spacing of settlements (R)

Threshold and range

The **threshold** is the minimum number of people required to support the provision of a good or service. The more specialised the service, the greater the threshold. For example, it has been estimated that a village shop requires a threshold of 500–1000 people, whereas a department store requires over 100 000.

The **range** is the maximum distance that people are willing to travel for a good or service. The range will depend on the value of the good or service, the time taken for the journey and the frequency of need. For example, people will not travel long distances for a newspaper (bought daily) but will for a piece of furniture (bought very infrequently). Therefore low-order services will have a small range and threshold and will be located close together, whereas high-order services will have a large range and threshold and will be located further apart.

You need to:
- be able to define threshold and range
- understand the relationship between threshold, range and the order of services

Spheres of influence

The **sphere of influence** of a settlement is the area served by the functions provided. In simple terms it is a summary of all of the ranges of the goods and services — economic, social and political — provided by that settlement. The sphere of influence tends to diminish with distance away from the settlement as fewer services are provided. The sphere of influence varies according to the order of the service provided, the proximity of competing centres, the density of the population, the physical nature of the terrain and administrative boundaries.

You need to:
- know what a sphere of influence is
- know the factors that cause variations in the sphere of influence
- know how to identify the sphere of influence

Reilly's law of retail trade gravitation

This model, based on the gravity model, is used to predict the theoretical margin (breakpoint) of the sphere of influence of competing centres. The calculations are performed using this formula:

$$D_j k = \frac{D_{ij}}{1 + \sqrt{\dfrac{P_i}{P_j}}}$$

where $D_j k$ is the breakpoint distance from the smaller town; D_{ij} is the distance between towns i and j; P_i is the population of the larger town; and P_j is the population of the smaller town.

By calculating and plotting the breakpoints for a number of centres of the same order around a settlement, the theoretical sphere of influence can be identified, assuming people travel to the nearest centre. This might not be the case, but the model, like central place theory, does provide a valuable vehicle for testing reality.

You need to:
- understand Reilly's law
- be able to calculate the theoretical sphere of influence and compare it with reality

Hierarchies

The settlement **hierarchy** refers to the arrangement of settlements by size or by impor-tance. A simple hierarchy would be hamlet, village, town, city. The increasing special-isation of settlements by order is seen in the central place theory.

The concept can also be applied to retail centres. You will be aware of the retail hierarchy (centres of different numbers and orders of shops) in your area from your study of retailing in the **Economic activity** section of Module 2 (see pages 42–52).

You need to:
- be able to define and give an example of a hierarchy
- be aware of the changes, and the reasons for the changes, in hierarchies from small (retailing within a town) to national scale (sizes of settlements)

Economic activity

Economic activity can be divided into several sectors. Definitions of the sub-divisions of the service sector are often inconsistent, and you will see different versions in textbooks. Those mentioned in each sector below are those used in the specification.

Sector	Definition
Primary	Economic activities which produce food, fuel and raw materials, e.g. agriculture, mining, quarrying, fisheries, forestry and primary energy production
Secondary	Economic activities which manufacture goods; these process raw materials, fabricate semi-finished products and assemble components, e.g. oil-refining, textiles, steel-shaping, electrical goods, motor vehicles etc.
Tertiary	Service activities, including retailing and offices

Over time, the importance of the sectors changes. Primary is the first sector to be important, followed, in sequence, by the others. As the tertiary and later sectors increase in importance, the earlier ones (primary and secondary) decline in relative importance. This is shown in the sector model, in which MEDCs have experienced the changes described above, whereas LEDCs have not yet experienced the full range of changes.

You need to:
- be able to define the sectors of economic activity
- know the relative importance of the sectors over time

Secondary activities (N)

Industrial change in the UK: de-industrialisation and growth

Industrial change has occurred in many parts of the UK. Some of the greatest changes have occurred in old coalfield areas, such as south Wales and northeast England. The heavy industries that were established on these coalfields to use the energy and raw materials, such as steel, shipbuilding and textiles, have been in decline for many years.

New manufacturing industries, producing consumer goods and being more market-oriented, were later established near to and in the main centres of population. These included the motor vehicle, electrical goods and food-processing industries. They were able to use road transport and electricity, so were not dependent on the coalfields.

Since 1960, **de-industrialisation** has been recognised in the UK. This is a rapid decline in the primary and/or secondary sectors of a country's or region's economy. There are two reasons for this decline: firstly, the shift from the primary and secondary sectors in MEDCs as the tertiary sector becomes more important; and, secondly, the growth in manufacturing in the LEDW as globalisation occurs.

Within the UK, as well as the decline in the primary sector, there has been a severe decline in a number of manufacturing industries — in particular, textiles and clothing, electrical and electronic goods, plastics and motor vehicles. The main reason for this decline is low-cost competition from abroad, especially Japan and parts of the LEDW, such as Singapore, Hong Kong and South Korea.

New industries have been established in the UK in recent years, including the electronics and computer industries. These have tended to cluster in certain parts of the country, such as southeast England (particularly along the M4 corridor), Scotland (Silicon Glen) and south Wales. The attraction of a well qualified labour force, proximity to universities for research, good national and international communications and a perceived pleasant environment have been important factors, as have government incentives in assisted areas (often in former coalfield areas, which have an available labour force).

You need to:
- understand the reasons for industrial change in the UK
- understand how and why de-industrialisation has occurred in the UK
- understand the reasons for industrial growth since 1960 in the UK

Industrial growth in the LEDW

Industrial growth has occurred in many countries in the LEDW. It is estimated that about 40% of the working population is in the formal sector, employed by the state or by multinationals (transnational) corporations. The remainder is in the informal sector. Large-scale, capital-intensive manufacturing industry is seen as important by governments, and incentives are provided to attract it.

It has been estimated that multinationals employ over 30 million people worldwide and the world's 100 largest multinationals control almost 50% of the world's manufacturing output.

The importance of multinationals to LEDCs and MEDCs

Multinationals make two locational choices: which country and which area within that country. The former involves an assessment of the operating environment, which depends on political factors.

In MEDCs, the attraction might be the aid package, involving cash grants, subsidies and tax concessions.

In LEDCs, the attraction might be a lack of operating restrictions. The benefits to the government and the country from the exploitation of resources, the use of capital and technology and the creation of employment outweigh any disadvantages of this.

In an MEDC, competition between regions results in the best overall financial package attracting the investment, though other factors, including the cost of land, communications and a skilled workforce, are important. For example, south Wales has attracted multinationals such as Sony, Panasonic and Ford. In an LEDC, the location is likely to be in the primate city, especially if it is the capital or the major port. The attraction of cheap labour, low-cost land and good communications are key factors in the decision.

The importance of multinationals to the economies of some LEDCs is demonstrated by newly industrialising countries (NICs), such as Hong Kong, Singapore, Taiwan and South Korea. Multinationals investing in these countries include those from MEDCs — in Hong Kong, for example, there are over 900 American multinationals — and those from the NICs themselves, with such household names as Daewoo, Proton and LG. Although the range of goods manufactured is very large, expansion is greatest in computing and IT. Sun Microsystems, Microsoft and SAP are important in Taiwan and Singapore.

You need to:
- be aware of the reasons for industrial growth in LEDCs
- recognise the importance of multinationals to this growth in NICs

The impact of these changes

There have been important economic, social, political and environmental impacts resulting from industrial growth involving multinationals in LEDCs.

Impact	Consequences
Economic	Employment is created for local labour; education, training and skills are gained; income is received by workers and governments; investment occurs; new technology is introduced; the multiplier effect stimulates local industry and services On the other hand, only a small number are employed relative to the size of the labour force; industry is capital-intensive so few jobs are created; wages are low; profits are exported; raw materials are exported instead of being processed locally; manufactured goods are exported rather than consumed locally; neo-colonialism (dependence on MEDCs)
Social	More employment opportunities; increased levels of education; better nutrition; healthier population; local provision of services improved, such as schools and health care On the other hand, long hours are worked; skills are kept at low levels; unemployment develops as mechanisation increases; the health and safety of the workforce are given low priority; conflicts arise between Western and local values
Political	Governments adopt policies to attract multinationals; acceptance of Western ideas; priority given to incomers at the expense of locals, causing resentment; governments gain prestige; lack of control in decision-making within the country; political and economic corruption can occur (crony capitalism); benefits gained only by the political elite; tendency towards dictatorship or a strong government under limited democracy
Environmental	Exploitation of resources causes environmental damage to landscape; deforestation occurs; pollution to rivers as industrial effluent is released; disease and malformed babies become more common

You need to:
- be aware of the role of multinationals in LEDCs
- be aware of the possible economic, social, political and environmental consequences of industrial growth in LEDCs

Variable industrial growth in LEDCs

The greatest industrialisation has occurred in the NICs. The reasons for this include:
- a well-educated, flexible and cheap labour force
- clear government support for industry
- free movement of imports and exports
- good accessibility and communication links

- ease of capital formation
- lack of restrictions on multinationals

At the other extreme, some LEDCs (particularly in Africa) have experienced very little industrialisation. Reasons include:

- war, drought and famine (all of which destroy the agricultural base)
- a lack of natural resources
- political instability
- corruption
- poor infrastructure
- dependency on the former colonial power

Most LEDCs fall in between these two extremes.

You need to:

- know how and why some LEDCs experience more industrialisation than others

Interdependence

Interdependence is the term that describes the relationship between MEDCs and LEDCs through trade, foreign investment, aid and migration. The relationship tends to favour the MEDCs, placing the LEDCs in a position of dependency. Aid is frequently tied — the money is advanced only for the purchase of specific goods from the donor. LEDCs export primary products, which have a low value, but they import high-value manufactured goods from the MEDCs ($650 billion to $735 billion).

Foreign investment from multinationals, however, has allowed manufacturing to be established in LEDCs.

Finally, it proves very difficult for the LEDC to break the links with the former colonial power.

The concept can also be applied to the operations of transnational corporations such as Ford and General Motors (Vauxhall) in the UK. Whole assembly plants have closed in Dagenham and Luton respectively, as a result of decisions taken at a global scale.

You need to:

- understand the term 'interdependence'
- be aware of the consequences of 'interdependence' for LEDCs and MEDCs

Tertiary and other activities (S)

Changing patterns of retailing in the UK

Retailing ranges in scale from the individual corner shop to the largest CBD and/or regional centre. A study of retailing in your local area is likely to provide a basis for further study and to demonstrate the patterns. Lower-order goods and services will be found in the smaller centres, such as corner shops and shopping parades close to residential areas. Higher-order goods and services will be found in larger, regional shopping centres, CBDs and regional centres.

You need to:
- be aware of the main changes to the pattern of retailing in the UK
- study these changes on a small scale in your local area

Changes in higher-order retailing

There has been increasing specialisation and concentration, both in the location of services and in terms of ownership. Three waves of decentralisation have been identified.

(1) Food

The growth in superstores and hypermarkets dates from the 1970s, shown by the expansion of the major supermarket chains. They were constructed in existing shopping areas, both in town centres and in suburban shopping centres. More recently, they have been built in out-of-town sites. The provision of car parking was an important element in their success as one-stop shopping became possible. There are now well over 8000 superstores in the UK.

(2) DIY and furniture

The second phase came in the 1980s, usually in the form of non-food retail parks which included large warehouses selling furniture, DIY, carpets, furnishings and gardening requirements. These were usually purpose-built, with a number of similar-sized outlets on each development. More recently, the retailing of clothing and electrical and computing goods has been increasingly located on such sites. Car access is important; these sites are often on main roads, sometimes occupying the sites released by the closure of older manufacturing industry.

(3) Retail shopping centres

The third wave came in the late 1980s and 1990s. The establishment of very large, regional retail centres was frequently the result of urban regeneration policies. These centres were built on very large sites, usually derelict land, for which there was no other obvious use. Examples include Dudley Merry Hill, which was built on the site of a former steel works, and Thurrock Lakeside (Essex) and Bluewater (Kent), which were constructed in former chalk quarries. There are five in all, the others being Sheffield Meadowhall and Gateshead MetroCentre. These centres rely on the motor car. (Bluewater has 13 000 free parking spaces, and a well-integrated public transport network with over 60 buses each hour and over 130 trains per day.) They are popular. Bluewater, for example, was visited by over 1 million people within 2 weeks of its opening in 1999. However, there are environmental concerns regarding their contribution towards dependence on the motor car.

It should be noted that, in response to the competition from out-of-town centres, most existing CBDs have been extended by the construction of a large retail facility, forming part of the existing town centre. These are usually on the fringes of the old CBD, thus allowing new car parking facilities to be included. Pedestrianisation of the shopping centre is frequently part of such a development.

You need to:
- know the three waves of decentralisation of retail services
- understand the reasons for these changes

Changes in lower-order retailing

There has been a rapid decline in small neighbourhood stores. These convenience shops, often family-run, have found it impossible to compete on price with the superstores, and require a specialised niche (such as personal service or quality) to survive. However, in recent years there has been a revolution in convenience stores, led by the establishment of large chains such as Alldays and 'metro' stores owned by the large supermarket chains. Not only is this type of store found in lower-order shopping centres, but also increasingly in petrol stations and in town centres.

You need to:
- be aware of the changes in lower-order retailing
- understand the reasons for such changes

The development of business and science parks

Business and science parks are purpose-built estates on the outskirts of cities or in regeneration areas with offices and factories for hi-tech businesses. The most common are associated with IT, computing, biotechnology and electronics.

One of the best known is at Cambridge, where research links with the university are of great importance. The Cambridge site is on the outskirts of the city, close to main road access. The land is owned by one of the colleges of the university, and planning permission was given as part of a policy to keep the city and the university at the forefront of technological advances.

Such developments are favoured by many authorities as there is little direct environmental impact on the surrounding area other than the use of greenfield sites. However, it is increasingly difficult to gain permission for development on greenfield sites.

You need to:
- understand the reasons for the establishment of business parks
- be aware of the characteristics of business parks
- understand the role of the planning process in the location of business parks

Out-of-town versus city centre

There are a number of issues that arise when out-of-town developments (for retailing and business science parks) are compared with those in the city centre. You will need to be aware of these as well as your attitudes, and those of others, towards such issues.

City centre disadvantages	Out-of-town disadvantages
Congestion Cost of parking car Decline in some services from competition Land for redevelopment more expensive Less space for development	Requires use of car for most people Excludes the poor and non-car owners Large amounts of land required Attracts custom from the city centre

City centre advantages	Out-of-town advantages
Good public transport access New shopping malls constructed Newly refurbished Pedestrianised More varied outlets	One-stop shopping Free parking Cheap land available Serves the wealthy Good accessibility Economies of scale Planned shopping environment Purpose-built modern buildings on business/science parks

You need to:
- be aware of the issues which arise
- explore your own attitudes and values to such issues
- explore the values and attitudes of others to such issues

Questions
&
Answers

This section of the guide contains three typical questions for Assessment Unit 2, based on the topic areas outlined in the Content Guidance section.

Sample answers are given after the questions. These are provided at a typical grade-C standard (Candidate A) and a good grade-A standard (Candidate B).

Examiner's comments

The examiner's comments are preceded by the icon *e*. They are interspersed in the answers and indicate where credit is due. In the weaker answers, they also point out areas for improvement, specific problems and common errors such as poor time management, lack of clarity, weak or non-existent development, irrelevance, mis-interpretation of the question and mistaken meanings of terms.

Most marks are awarded according to 'Level' attained. For each question:
- part (a) is points marked
- part (b) has two Levels:
    - Level 1 (basic) 1–3 marks
    - Level 2 (clear) 4–5 marks
- part (c) has three Levels:
    - Level 1 (basic) 1–4 marks
    - Level 2 (clear) 5–7 marks
    - Level 2 (detailed) 8–10 marks

Population dynamics

(a) Outline the characteristics of the demographic transition model. (5 marks)
(b) How useful is the demographic transition model? (5 marks)
(c) Examine the links between population structure and the demographic transition model. (10 marks)

■ ■ ■

Answer to question 1: Candidate A

(a) The demographic transition model is described by the following diagram.

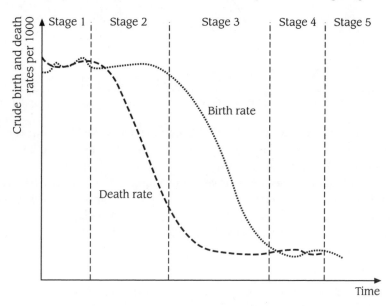

The stages can be seen clearly. Stage 1 has high birth and death rates caused by a lack of birth control and disease. Stage 2 has falling death rates and high birth rates, caused by better nutrition, health care etc. Stage 3 has falling birth rates to meet the death rates, caused by changing roles of women in society. Stage 4 has low birth and death rates. Stage 5 might exist if birth rates fall below death rates.

> ⓔ This response lacks focus. The diagram of the model lacks the total population or population growth line and the stages are not specifically named. In addition, no figures are quoted for the stages. 2 marks would be awarded.

(b) The model is very useful as it allows all countries to be placed on it and to be compared with others. It is, however, based on European countries and it is not certain whether all countries, especially those in the LEDW, will follow this pattern.

Many countries are experiencing population growth without the economic development which MEDCs followed, so the outcomes might be different. The model is useful for research on population growth.

 This response is quite perceptive and covers the usefulness well. There is enough here for 3 marks (the top of Level 1).

(c) The links between population structure and the DTM can be seen in terms of the population pyramids. Each stage has its own typical pyramid, as shown in the diagram below.

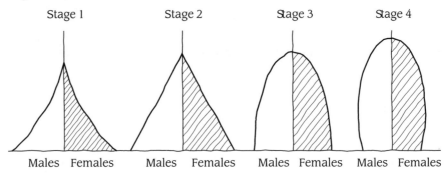

Stage 1 has an expansive pyramid which is the result of the high birth and death rates. The life expectancy is also low, so the pyramid is short. Stage 2 has linked to it another type of expansive pyramid, but as the death rate has fallen, it has straight sides, rather than curved ones. Stage 3 has a stationary pyramid, identified as a result of the fall in both death and now birth rates. This has more vertical sides as, of the numbers born, very few die. Life expectancy is higher, so the pyramid is taller. Stage 4 is the contractive pyramid. As the birth rates fall and death rates are low, the base is smaller and there is a long life expectancy.

Therefore, there is a link between the DTM and the structure of a population.

 Although the main links are made and life expectancy is covered, there is a lack of detail in that the stages of the DTM are not specifically named. Similarly, there is no reference to the concave or convex shapes of the pyramids. An essay format is not always apparent, but there is a conclusion in this case. This is a Level 2 response, for 5 marks.

■ ■ ■

Answer to question 1: Candidate B

(a) The demographic transition model shows the effects that changes in crude birth and death rates have on the population of a country. Its characteristics are best shown in the following diagram.

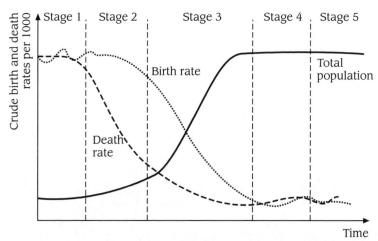

Stage 1, the high fluctuating stage, has high birth and death rates, over 35/1000, giving a slowly growing population. Stage 2, the early expanding stage, has declining death rates (20/1000), which with the still high birth rates gives a rapidly increasing population. Stage 3, the late expanding stage, has low death rates (15/1000) and declining birth rates (20/1000), causing population growth to start slowing down during this stage. Stage 4, the low fluctuating stage, has low birth and death rates (16/1000 and 12/1000 respectively), and by now the population is high in comparison with the starting one. Some have indicated a Stage 5, wherein birth rates fall below death rates to give a declining population, but others see this as merely the fluctuations in Stage 4.

✓ This is virtually the perfect answer, for full marks. There is a full, detailed description of the characteristics of the model, including reference to specific figures.

(b) The model is useful in several ways. It has universal application, in that all countries can be applied to the stages of the model, and certain countries seem to fit well. In particular, European countries, such as the UK, appear to have passed through all of the stages. It can be seen to be predictive in that it is possible to foresee the next stage for a particular country. However, not all countries, especially those in the LEDW, will follow the model because of other factors such as famine or government intervention. It also has a role as a starting point for further research and as a comparator with other countries. Therefore the model has a number of uses for the study of population geography.

✓ A competent summary of the model's usefulness is set out, with positive and negative comments. Again, full marks are awarded.

(c) There are very clear links between the DTM and population structure. This can be most clearly identified by looking at the four stages of the model, which also correspond to typical population pyramids showing population structure.

question

Stage 1, the high fluctuating stage, relates to the expansive pyramid, with its low, concave profile, indicating high birth and death rates and short life expectancy.

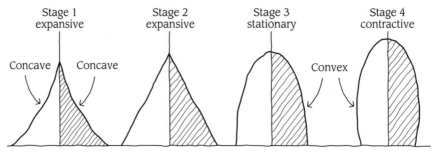

Stage 2, the early expanding stage, relates to the second expansive pyramid, indicating a fall in the death rate which means the pyramid loses its concave shape as fewer people are dying. The pyramid is also taller as life expectancy increases.

Stage 3, the late expanding stage, relates, in turn, to the stationary pyramid, whereby the declining birth and death rates cause a rapid fall in the numbers in the younger age groups and fewer of the population die. As a consequence, the pyramid now starts to have a convex profile. The life expectancy is still increasing.

Stage 4, the low fluctuating stage, relates to the contractive pyramid. The low birth and death rates cause a more strongly convex pyramid as the effects of the low birth rate impact on the numbers born. This pyramid indicates the highest life expectancy.

Therefore, these ideal pyramids (referring to the stages of the model) show how the population structure of a country changes over time as the sequence develops. The links are clearly made and seen.

📝 This response meets the demands of the question and is awarded full marks. The stages of the DTM are linked effectively to the ideal pyramids. The characteristic shapes of the pyramids are shown to be a product of the birth and death rates at each stage of the model. Thus, the transition of population structure which each country experiences is identified. There is also a clear essay format.

Settlement processes and patterns

(a) Define the term 'million city' and explain how the location of these cities has
changed over time. (5 marks)

(b) Describe the consequences of rapid urbanisation in one LEDC. (5 marks)

(c) Assess the relative contribution of brownfield and greenfield sites to meeting
the need for additional housing. (10 marks)

Answer to question 2: Candidate A

(a) A million city is one with a population of over 1 million inhabitants. Over time,
there have been changes in their location. In the industrial revolution, many large
cities grew up and in recent years new ones have grown in the LEDW. Most of
them are now found in this area, for example Shanghai and Calcutta.

> The definition is adequate and accurate. However, the locational change is less well
> covered, being generalised, apart from the examples mentioned at the end. This
> response would be awarded 3 marks out of the 5 (a mark each for the definition,
> the mention of a change in location and the citing of an example).

(b) Rapid urbanisation in Brazil has many consequences. The main one is the building
of shanty towns or *favelas*. These are made of cardboard, tin or any building
material that can be found by the poor of the city. There is no electricity or gas
and running water may be provided by means of standpipes. The people who live
here are poor and unemployed. As there is no running water or sewers, disease
is common.

> This is a general outline of some of the main consequences of urbanisation. The need
> for a case study is barely met, with just a brief mention at the start. Thus, no detail
> is provided. This would reach the top of Level 1 and gain 3 marks.

(c) There is a need for more housing to be built in the UK and the government wants
to use brownfield sites rather than greenfield ones. Brownfield sites are found in
urban areas, usually on land reclaimed from other uses, such as industry or landfill.
As such, this is a good use of land, but not many people want to live in these
areas. Greenfield sites are more popular, as they are clean and unpolluted. They
are located in the rural areas around the cities and do not need to be reclaimed
for building. They are built at a lower density and require new services, such as
water and schools, to be added, thus increasing the costs. In summary, brownfield sites are a better use of land, but people would rather live in greenfield areas.

> This is a relatively brief outline of the two types of site. There are some charac-
> teristics of both types covered, but the command word 'assess' is implicitly, rather
> than explicitly, developed. Detail is lacking, as no examples are quoted. Overall, this
> would remain in Level 1 and be awarded 4 marks.

Answer to question 2: Candidate B

(a) A million city is defined as a city with 1 million or more people living in it. The first million cities were established in the areas of the industrial revolution in the MEDW. London was one of the first and by 1951 was the largest city in the world with a population of 8 million. Other examples included Paris, Chicago and New York. At this time, and up until 1950, million cities were found in mostly temperate latitudes of between 40 and 60°N. After 1950, the growth came in low latitudes in the LEDW as rapid population growth and inmigration caused many capital cities, such as Mexico City, Calcutta and Jakarta, in the LEDW to expand rapidly. Thus, the locational shift was from high to low latitudes.

> ✍ The definition is clear and there are examples from both worlds to demonstrate the locational shift. Timescale is included in the answer. This would gain the full 5 marks.

(b) Rapid urbanisation has resulted in a number of consequences in urban areas in the LEDW. In Brazil, for example, the growth of shanty towns, or *favelas*, has been consistent over the last 40 years or so. In São Paulo, the *favelas* have grown on the outskirts of the city and on any spare land within it, including steep slopes. Shanties are, to an extent, a consequence of population growth and migration. The population of these *favelas* is usually poor. They build their dwellings out of any available materials. Water and sewerage are limited and pollution of water supplies by sewage and industry causes disease to be rife. Supplies of electricity are variable. There are high levels of unemployment and underemployment, despite the willingness of the people to work. One consequence of these conditions is that the government is driven to intervene. In São Paulo, this has been by the provision of site and service schemes. Water and power are provided and the people build their own properties on the sites made available, thus giving them control of their environment to some extent.

> ✍ This is a full response. The characteristics are well described, with some support. Detail is present and the range of points is extensive, covering the governmental responses that change the characteristics of the *favelas*. This would be clearly at the top of Level 2 and gain full marks.

(c) Brownfield sites are those found in urban areas, reclaimed from industrial or extractive uses. Their use for housing is important as the land is already in urban areas, and services, such as electricity and water, are more straightforward to connect. The government needs to allow the use of such generally smaller sites to meet its housing targets. Frequently, the sites are contaminated, so there are high costs attached to reclamation, and housing may be blighted by the pollution. Areas of the London Docklands would be good examples, with housing developments in Surrey Quays, Whitechapel and Beckton built or converted at high density.

On the other hand, greenfield sites are preferred by most people. They are on the outskirts, such as the large-scale development around Ashford, Kent. The environment is cleaner and more attractive and densities are lower. However, the cost is higher, as it takes a longer time to gain planning permission and there is usually loss of agricultural land or woodland. New infrastructure has to be provided to enable water, power and gas to be connected, and new schools and health facilities have to be constructed.

Overall, to meet targets, the government has to use both brownfield and greenfield sites. The former are preferred but will not by themselves meet the demand for housing, especially in the south of England, without the use of greenfield sites.

☑ The command words are followed well and there is a consideration of the strengths and weaknesses of each type of site. In addition, there is an evaluation of their contribution to meeting the targets set by the government. Examples are used to support points. There is also a good style of expression. This is a clear Level 3 response and could gain full marks.

uestion 3

Economic activity

The rationalisation of the global car industry is continuing into the twenty-first century. It is anticipated that within 5 years there will be only five major-volume car manufacturers. Some groupings are shown in the following table.

Group	Brands/links
Ford	Ford, Jaguar, Lotus, Mazda, Volvo, Range Rover
General Motors	Vauxhall, Opel, Saab
Daimler-Chrysler	Mercedes, Chrysler
Volkswagen	Volkswagen, Seat, Skoda
Renault	Renault, Nissan, Samsung
Peugeot-Citroën	Peugeot, Citroën
BMW	BMW, MG

(a) What does the information above indicate about recent industrial change in the motor vehicle industry? (5 marks)

(b) What changes in location of manufacturing plants are likely to occur as multinational companies increasingly dominate production? (5 marks)

(c) Explain some of the social and economic impacts of these changes. (10 marks)

Answer to question 3: Candidate A

(a) There are several major car-making groups. They own many other car makes from different parts of the world to enable them to get bigger and compete. For example, there are American and European car-makers in this list of companies which have taken over other firms in both Europe and Japan. It is likely that there will be fewer car-makers in the future.

> This is a fairly basic response, for 2 marks. The general principles are here, with reference to the consolidation and origins of the major groups, but there is no clear reference to the role of a multinational company or the full range of companies in the major groups. Future trends are mentioned.

(b) As there are fewer car firms, there will be fewer plants to make them in. The firms will want to make cars and to transport them to the market as cheaply as possible. This means that, in the long run, there might be fewer cars made in Europe and these car plants are more expensive to run in comparison with those in the LEDCs. On the other hand, the main markets are in Europe, so some plants might be kept here.

📝 The principles are here, but in outline. The shift to the best site is covered, as is the idea that there will be fewer plants. Increasing specialisation, though implied, is not mentioned. The idea of location in LEDCs is also mentioned, but the market arguments remain inconclusive. The response omits reference to changes in the location of component plants. This is a Level 1 answer, gaining 2 marks.

(c) When there is a change of location for a car plant, there are large numbers of people involved, both in the factory itself and among component manufacturers. If there is a new factory built, then many jobs are created, which benefits the surrounding area. The government gets the prestige of this investment. The component manufacturers also benefit from the demand created. Socially, there are gains as there are fewer problems in the area.

📝 This is a very unbalanced response, placing it at the top of Level 1, for 4 marks. The comments on the economic advantages are, in general, competent, though generalised as there is no specific support. The lack of detail or depth on the social impacts is the main weakness of this response.

■ ■ ■

Answer to question 3: Candidate B

(a) There are currently seven major, multinational, volume car manufacturers operating at a global scale. This is a result of consolidation, i.e. the take-over of other car manufacturers to gain economies of scale. Of those in the table, two are American (Ford and General Motors), and the others European (Daimler-Chrysler, Volkswagen, Renault, Peugeot-Citroën and BMW). The companies taken over are from both MEDCs and LEDCs as a result of globalisation. The independence of the two smaller groups might be short-lived as consolidation continues.

📝 This response comments on major changes which have affected the motor industry: consolidation, globalisation and increasing multinational ownership and operations. Other points refer to the location of the group, and the links between the LEDW and MEDW. Future trends are projected from the recent trends. This is a more than competent answer, for full marks.

(b) Multinational or transnational companies operate at a global scale. Therefore, there are likely to be fewer, but more specialised, manufacturing plants in order to make economies of scale to produce and supply the market with the cars produced. This will mean that cars and components will be manufactured at the best site in terms of cost, accessibility and transport facilities. An example is the decision of Ford to cease car production at Dagenham and supply the UK market from Belgium and Germany. At the same time, the Dagenham site will be a major engine producer for Ford, thus emphasising the specialisation in components.

question

📝 The candidate identifies the major changes that are likely — fewer factories and at the best sites. The case study is particularly relevant and is a good example of the support to be gained from an up-to-date knowledge of current affairs. The link with trends in components is also very relevant. Full marks are awarded.

(c) Because of the sheer scale of such plants, any change of location will result in great social and economic impacts. The example of Ford at Dagenham is a case in point, where over 3000 jobs will go. There are immediate job losses if a plant closes down, both directly and in component suppliers — several thousand in this case. There will also be a downward multiplier in the surrounding economy if and until those made redundant find other employment, thus resulting in further job losses. The economy of the country will suffer if measures are not taken to ameliorate the position. There are also prestige considerations, as car manufacturing is given a high status. The current government announced retraining packages for workers losing their jobs. Social impacts are considerable, including potential long-term unemployment, the consequent impact on family life and a higher incidence of health problems. All of these will require intervention and support from local and national government agencies.

📝 This is a full response, and is stronger on the economic than the social impacts. The imbalance is outweighed, however, by the quality of the points and the fact that they are supported by up-to-date case material. The answer does concentrate on the loss of a plant, but it would be valid to look at the benefits in an area where such a plant is located, or to balance the two. The important point for this grade is that both social and economic impacts must be covered. The use of language (English and geographical) confirms this as a Level 3 response, for full marks.